YOUR KNOWLEDGE HAS VALUE

- We will publish your bachelor's and master's thesis, essays and papers

- Your own eBook and book - sold worldwide in all relevant shops

- Earn money with each sale

Upload your text at www.GRIN.com and publish for free

Amalia Aventurin

Gas Breakthrough

Petrophysics

GRIN Verlag

Bibliografische Information der Deutschen Nationalbibliothek:

Die Deutsche Bibliothek verzeichnet diese Publikation in der Deutschen National-
bibliografie; detaillierte bibliografische Daten sind im Internet über http://dnb.d-
nb.de/ abrufbar.

Imprint:

Copyright © 2013 GRIN Verlag GmbH
Druck und Bindung: Books on Demand GmbH, Norderstedt Germany
ISBN: 978-3-656-64486-6

This book at GRIN:

http://www.grin.com/en/e-book/272606/gas-breakthrough

Gas Breakthrough

I. Introduction: Principle Measurement

By inducing a gas (non-wetting fluid) into a water saturated rock, the gas will displace the water (wetting fluid), but this process just take place, when the capillary pressure is above the capillary entry pressure (gas pressure difference). First the largest pores near the sample surface are drained (drainage process). At higher capillary pressures even the smallest pores are filled with gasThe relation between capillary pressure and pore radius is given by the Washburn equation (1921), "The intrusion of a non-wetting fluid into a cylindrical capillary of radius r only occurs if the capillary pressure P_c [...] within a pore is exceeded" [1]:

$$P_c = P_1 - P_2 = \frac{2\gamma \cdot \cos\theta}{r} \tag{1}$$

Where P_c is the capillary pressure, γ is the interfacial tension, r the pore radius and θ the wetting angle of the fluid in the capillary.

At gas breakthrough gas starts to flow through the sample and displace water from the pore system (drainage path). The effective permeability of the gas-phase after the gas breakthrough is a function of the gas/water saturation and is no longer a rock property like the absolute permeability (single-phase flow). The effective permeability can be determined using the following equation:

$$k_{eff} = -\frac{V_2 \cdot \eta \cdot 2\Delta x}{A(P_2^2 - P_1^1)} \cdot \frac{dP_2}{dt} \tag{2}$$

$$with \; \frac{dV}{dt} \cdot P_2 = \frac{dn}{dt} = mass\; flow \tag{3}$$

A decrease of capillary pressure leads to an imbibition of water and gas saturation decreases until the last capillary pore is blockaded and gas flow is restricted (snap-off pressure). Complete pressure equilibrium is not established.

As the experiments were conducted with gas Darcy's law for compressible media is used and the data are Klinkenberg corrected. This is done to determine the absolute gas permeability. At low mean fluid pressures, the slip flow effect causes a non-zero velocity at the pore walls. As a consequence the average flow velocity is higher. At infinite high fluid pressures the mean free path length of the gas molecules is strongly reduced and fluid flow is comparable to those of liquids.

The Klinkenberg correction is done by the following equation:

$$k = k_\infty \cdot \left(1 + \frac{b}{p_{mean}}\right) [\frac{m^2}{s}] \tag{4}$$

A correction has not be done for liquid fluids, as here a normal velocity profile is given (zero at the pore walls).

Two different gas breakthrough experiments were done. The first was performed under steady-state conditions (constant pressure and constant volume flux) as shown in figure 1, using a dried and saturated sample. The experiment on the dried sample represents a single-phase flow. The experiment on the saturated sample represents a two-phase flow after the gas breakthrough occurred and gas displaced the water phase. The second experiment was conducted under non-steady-state conditions. Here a high pressure difference was created in the beginning of the experiment and the imbibition process was measured with final detection of the snap-off pressure (final pressure difference).

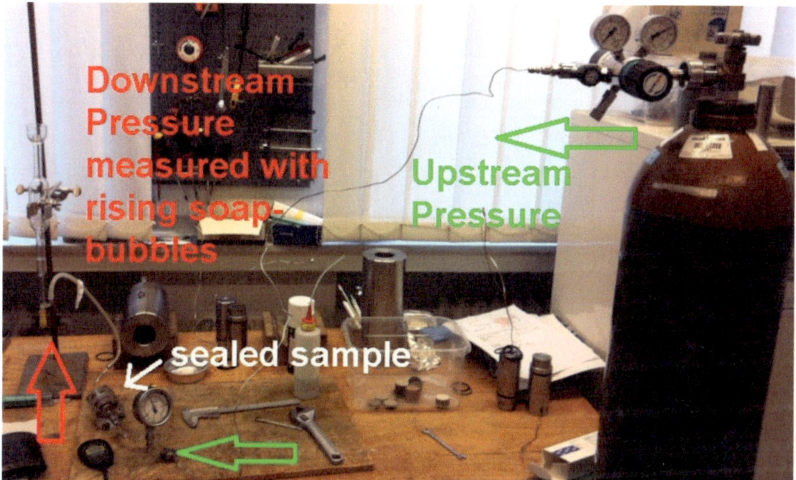

Fig. 1: Apparatus for the single-phase gas and water permeability experiment.

Steady state experiment

In the first experiment, the sample is placed in a PE-cell. No confining pressure is applied. Bypass of the gas is prevented by an O-ring. One side held constant at atmospheric pressure, the other side with increasing p1-values due to induced helium. For the dried sample a Klinkenberg correction has to be done to calculate the absolute permeability, which is not necessary for the saturated sample. Here the result is a desaturation of the system and therefore we get the effective permeability. Gas flow rates are determined on the low pressure side using a bubble flow meter.

Non-steady state experiment

Before the gas breakthrough experiment could be started, a calibration was performed (figure 2) in order to receive the volumes of the upstream and downstream pressure cell. A schematic overview of the gas breakthrough is given in figure 3.

Calibration:

To calibrate the sample cup there has to be three values constant. These values are temperature (T), Railey-constant (R) and (n) as the amount of substance. The values which get measured are volume and pressure and applied in the ideal gas law:

$$p \cdot V = n \cdot R \cdot T \tag{5}$$

Out of the ideal gas law the formula to calculate the unknown volume (V_X) in the cell:

$$\Delta p_R \cdot V_R = \Delta n_R \cdot R \cdot T$$

$$\Delta p_X \cdot V_X = \Delta n_X \cdot R \cdot T$$

$$\Delta n_X = \Delta n_R$$

$$\Delta p_R \cdot V_R = \Delta p_X \cdot V_X \rightarrow V_R \cdot (p_{R1} - p_{R2}) = V_x \cdot (p_{x2} - p_{x0})$$

$$V_X = \frac{p_{R1} - p_{R2}}{p_{X2} - p_{X0}} \cdot V_R \tag{6}$$

With V_R as reference volume.

Four different calibration steps has to be done by closing four different valves in different ways. These four steps get repeated with the valves in the volume calibration set up five times in each step five repetitions. If this is done the sample in the sample cup get measured during a few hours. It's a closed system without a constant flux, in this case the induced pressure (different in the upper and bottom part) comes in balance. Depending on how fast the flux is you can measure the permeability.

Imbibition experiment:

This experiment is under non-steady state condition. It's a closed system without a constant flux, in this case the induced pressures (different in the uppor and bottom part) start to equilibrate. From the flux (pressure drop, increase in closed volumes V1, V2) the permeability is calculated.

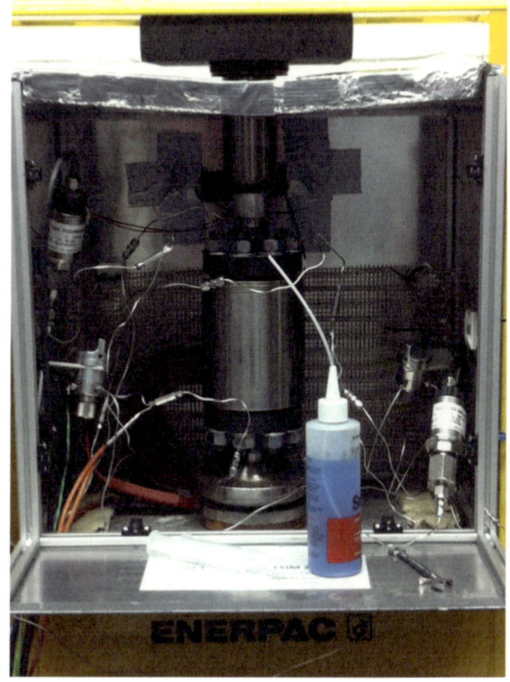

Fig. 2: Used apparatus for the gas breakthrough experiment.

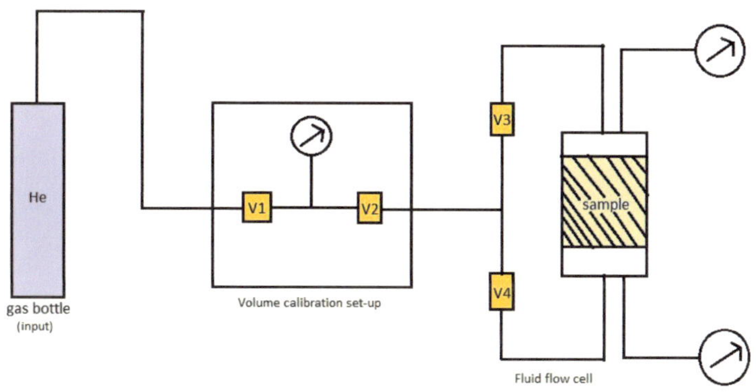

Fig. 3: Schematic overview of the gas brakthrough apparatus.

The results for the calibration are shown in table 1.

Measured dead	Measured Dead+Top	Measured Dead + Bottom	Calc Top	Calc Bottom	Calc Sum	Measured All	Diff (all-Sum)
1.7E-06	$4.2 \cdot 10^{-6}$	$3.8 \cdot 10^{-6}$	$2.5 \cdot 10^{-6}$	$2.1 \cdot 10^{-6}$	$6.2 \cdot 10^{-6}$	$6.2 \cdot 10^{-6}$	$-2.4 \cdot 10^{-8}$

Tab. 1: Calibration results for non-steady state single-phase flow experiment.

II. Results steady state experiment

II.1. Steady-State single-phase gas permeability on a dry sample

A dried sandstone sample with a length of 0.02185 m and a diameter of 0.0283 m and a cross section of 0.000629018 m² was used, with an accuracy of the callipper about 0.00005 m. The measured results are shown in table 2. The used µ-value is 0.0000195 Pa/s.

t [s]	P_1 relative [bar]	dV (bubbles) [m³]	dV/dt	P_1 absolute [Pa]	P_1 relative [Pa]
43,54	0,435	0,000001	2,29674E-08	144825	43500
44,97	0,355	0,000001	2,2237E-08	136825	35500
44,16	0,35	0,000001	2,26449E-08	136325	35000
23,85	0,56	0,000001	4,19287E-08	157325	56000
18,55	0,665	0,000001	5,39084E-08	167825	66500
16,41	0,7475	0,000001	6,09385E-08	176075	74750
13,87	0,83	0,000001	7,20981E-08	184325	83000
12,62	0,875	0,000001	7,92393E-08	188825	87500
11,09	0,955	0,000001	9,01713E-08	196825	95500
9,68	1,05	0,000001	1,03306E-07	206325	105000
111,18	1,065	0,000001	8,09498E-08	207825	106500
104,34	1,05	0,000001	8,62565E-08	206325	105000
90,84	1,15	0,000001	9,90753E-08	216325	115000
82,66	1,255	0,000001	1,0888E-07	226825	125500
72,47	1,355	0,000001	1,24189E-07	236825	135500
68,47	1,45	0,000001	1,31444E-07	246325	145000
61,91	1,525	0,000001	1,45372E-07	253825	152500
58,59	1,6	0,000001	1,5361E-07	261325	160000
54,5	1,775	0,000001	1,65138E-07	278825	177500
51,72	1,8	0,000001	1,74014E-07	281325	180000
48,15	1,925	0,000001	1,86916E-07	293825	192500
42,44	2,15	0,000001	2,12064E-07	316325	215000

Tab. 2: Measurement and calculation results for sample 29.3.

The calculated permeability coefficients (k) are given in table 3. The plotted Klinkenberg diagram with trend line is shown in figure 3a. The gas breakthrough event is shown in figure 3b.

k	Q	$(dV/dt)\cdot P_2$	P_{mean} [bar]	$1/P_{mean}$ [bar]
2,94437E-16	3,65131E-05	0,00232717	1,23075	0,812512696
3,61051E-16	3,5352E-05	0,00225317	1,19075	0,839806844
3,7371E-16	3,60005E-05	0,0022945	1,18825	0,841573743
3,97357E-16	6,66575E-05	0,00424843	1,29325	0,773245699
4,13438E-16	8,57025E-05	0,00546226	1,34575	0,743080067
4,03407E-16	9,68788E-05	0,00617459	1,387	0,720980534
4,17428E-16	0,00011462	0,00730534	1,42825	0,700157535
4,2843E-16	0,000125973	0,00802892	1,45075	0,689298639
4,34711E-16	0,000143353	0,00913661	1,49075	0,670803287
4,38984E-16	0,000164234	0,01046746	1,53825	0,650089387
3,37495E-16	0,000128692	0,00820224	1,54575	0,646935145
3,66535E-16	0,000137129	0,00873994	1,53825	0,650089387
3,72296E-16	0,000157508	0,0100388	1,58825	0,6296238
3,62912E-16	0,000173095	0,01103224	1,64075	0,609477373
3,72054E-16	0,000197434	0,01258348	1,69075	0,591453497
3,57933E-16	0,000208968	0,01331861	1,73825	0,575291241
3,68443E-16	0,00023111	0,01472985	1,77575	0,563142334
3,63397E-16	0,000244206	0,01556452	1,81325	0,551495933
3,35941E-16	0,000262533	0,01673257	1,90075	0,526108115
3,468E-16	0,000276644	0,01763196	1,91325	0,522670848
3,37305E-16	0,000297155	0,01893925	1,97575	0,50613691
3,2418E-16	0,000337135	0,02148739	2,08825	0,478869867

Tab. 3: Calculated results for the sandstone sample 29.3by using Darcy's law for compressible media.

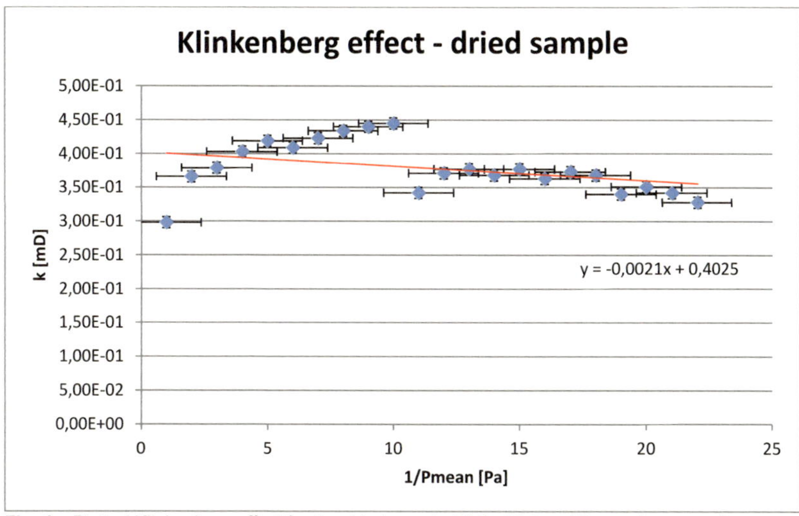

Fig. 4a: Plotted Klinkenberg effect for the dried sample 29.3 with trend line and error scale.

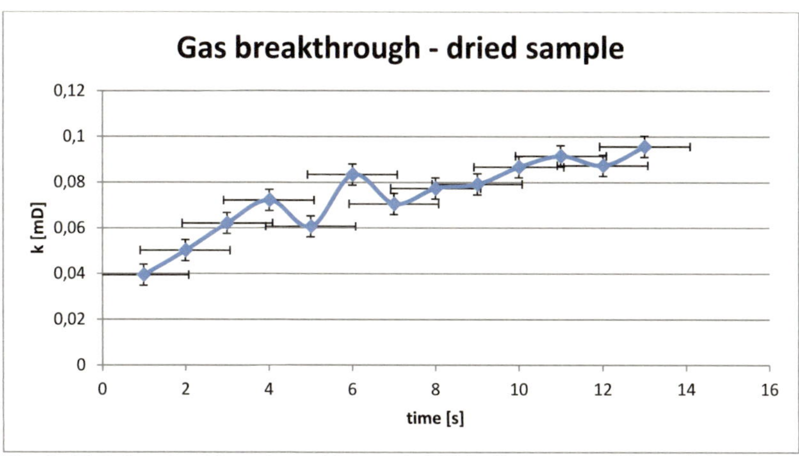

Fig. 4b: Plotted gas breakthrough event for sample PR0-1 with the permeability coefficient (k) over time. Including error bars and trend line.

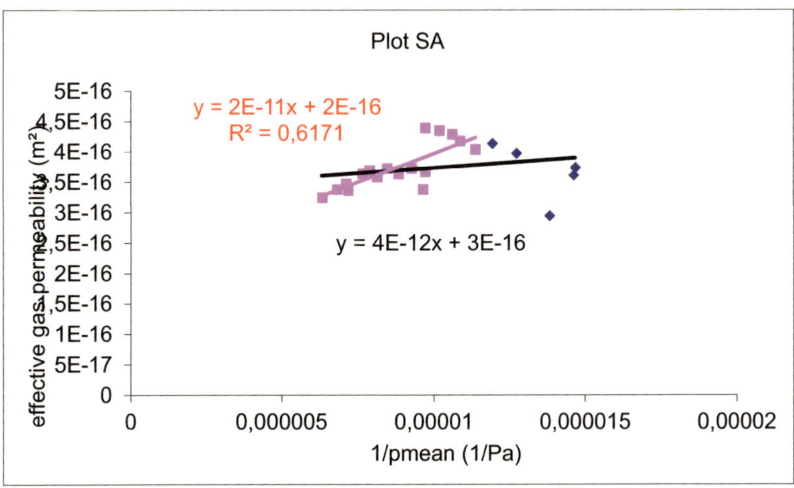

Fig. 4c: effective gas permeability [m²] over the reciprocal mean pressure.

As in figure 4a shown, an incorrect Klinkenberg effect is seen for this sample. The error can be caused by a false experimental procedure by closing the gas valve after the gas pressure was set. Figure 4b is just an additional diagram to plot the permeability over time. It can be observed that with increasing gas influx the permeability coefficient (k) increases, because more and more pores are filled with helium. Figure 4c shows the effective gas permeability [m²] plotted over the reciprocal mean pressure.

II. 2. Steady-State gas permeability on a saturated sample

This gas breakthrough and two-phase water permeability experiment was done on a saturated sandstone sample, with a length of 0.0206 m and a diameter of 0.02835 m and a cross section of 0.000631242 m². The size determination has again an error of about 0.00005 m. By using the results and Darcy's law the permeability coefficient (k) can be calculated. The measurement and calculation results are shown in table 4. The used µ-value is 0.0000195 Pa/s. The gas breakthrough event is shown in figure 4b. A Klinkenberg correction is not neceassary for this experiment. Just to compare both samples and their results a Klinkenberg plot can be done.

t [s]	P₁ abs [Pa]	P₁ rel [bar]	dV (bubbles) [m³]	(dV/dt)·P₂	K [mD]	Q
-	201325	1	0,000001	-	-	-
46,22	286325	1,85	0,000001	46,22	3,89055E-17	3,42747E-05
19,84	376325	2,75	0,000001	19,84	4,94844E-17	7,98477E-05
10,15	466325	3,65	0,000001	10,15	6,13215E-17	0,000156077
6,16	551325	4,5	0,000001	6,16	7,12818E-17	0,000257172
52	651325	5,5	0,000001	5,2	5,99092E-17	0,00030465
30,28	726325	6,25	0,000001	3,028	8,23322E-17	0,000523176
61,15	201325	1	0,000001	61,15	6,96811E-17	2,59064E-05
20,97	301325	2	0,000001	20,97	7,63653E-17	7,5545E-05
10,94	401325	3	0,000001	10,94	7,81715E-17	0,000144806
6,25	501325	4	0,000001	6,25	8,55948E-17	0,000253468
40,65	601325	5	0,000001	4,065	9,0299E-17	0,000389712
31,06	701325	6	0,000001	3,106	8,62132E-17	0,000510038
21,62	801325	7	0,000001	2,162	9,44016E-17	0,000732737

Tab. 4: Measurement and calculation results for sample PR0-1 using Darcy's law for compressible media.

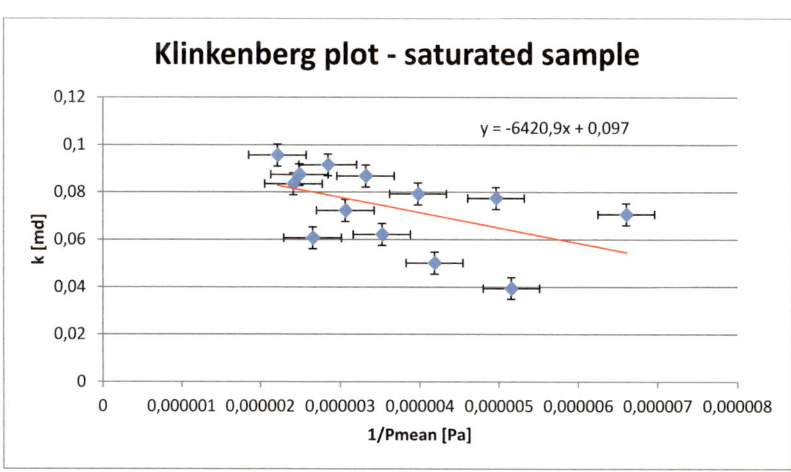

Fig. 5a: Plotted Klinkenberg effect for the saturated sample, including error bars and trend line.

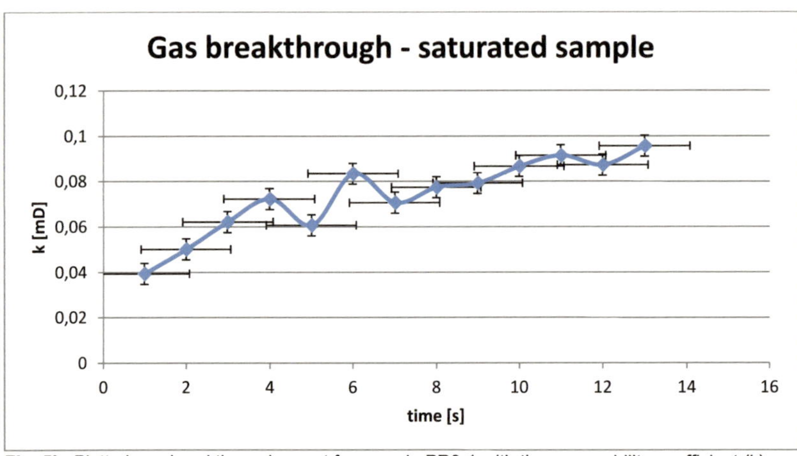

Fig. 5b: Plotted gas breakthrough event for sample PR0-1 with the permeability coefficient (k) over time. Including error bars and trend line.

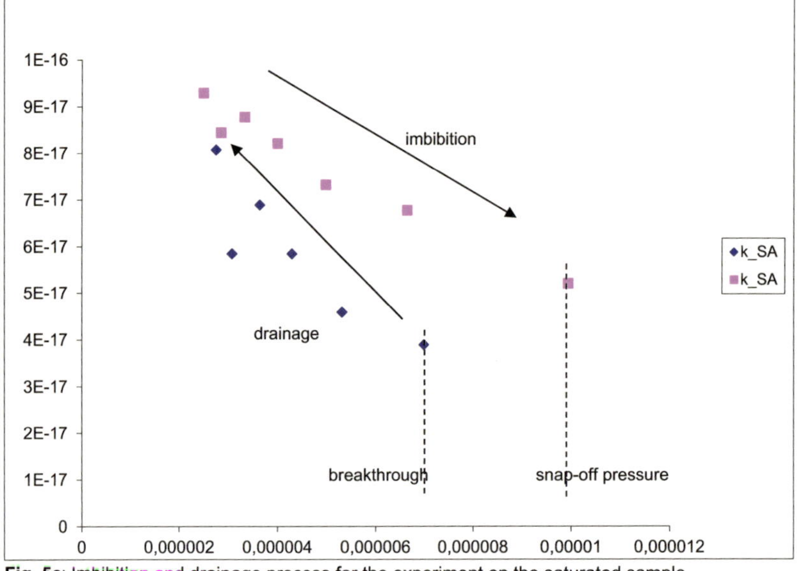

Fig. 5c: Imbibition and drainage process for the experiment on the saturated sample.

Figure 5a shows a Klinkenberg plot for the used sample for a comparison with the dried sample. For this rock a Klinkenberg effect was not observable. Figure 5b shows again an increase of the permeability coefficient (k) over time. This caused by the effect that helium replaces the water (drainage). Figure 5c shows the observed imbibition and drainage process for the saturated sample.

II. 3. Gas Breakthrough experiment

The input parameters to calculate the permeability are shown in table 5.

Input Data	Values
Volume top cell (V_1)	0.0000021 m³
Volume bottom cell (V_2)	0.0000025 m³
Sample thickness (dx)	0.01116 m
Diameter (D)	0.02745 m
Helium viscosity (μ)	0.0000191596573564114 Pa·s
Cross section area (A)	0.00059 m²
Helium density (ρ)	0.1785 kg/m³
Pressure transducer (p_1)	40000000 Pa (average 4.0E+06 Pa)
Pressure transducer (p_1)	25000000 Pa (average 3.4E+06 Pa)

Tab. 5: Input parameter for the measured sample.

The estimated errors are low, because all values originate from calibrated and standardized machines, and are by about ± 0.0000001 %. The estimated relative maximum errors for both pressure transducers are 0.05 %.

Figure 6 and 7 show the results for the gas breakthrough experiment. In figure 6 can be observed that the gas breakthrough occurs nearly at the beginning of the experiment by around 50 seconds (estimated). After imbibition a residual snap-off pressure remains. Figure 7 shows the gas breakthrough and snap-off once again and additionally the effective permeability (k_1 and k_2). It can be observed, that k_1 decreases and k_2 increases until gas breakthrough occurred. After this the imbibition of water is started, both k_1 and k_2 decrease until just the snap-off pressure remains, than both values are quite constant. The average of the used pressure values are represented in table 5. The average of the used permeability coefficient (k) are: $k_1 = 1.1·10^{-21}$ m² and $k_2 = 3.1·10^{-21}$ m².

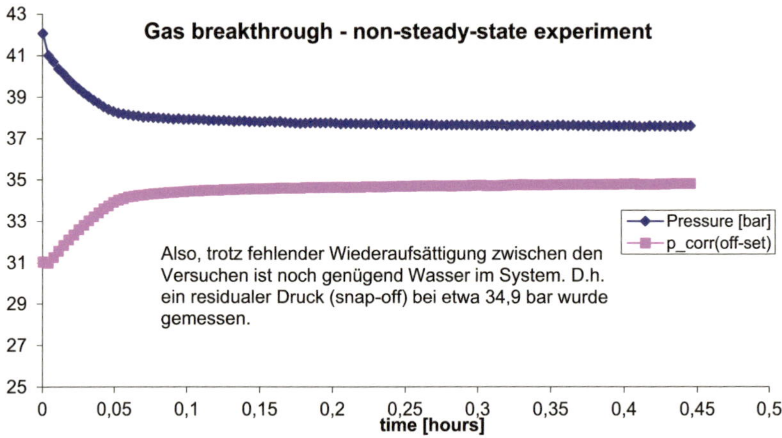

Fig. 6: Gas breakthrough experiment under non-steady-state conditions. On the y-axis is the fluid pressure [bar] plotted against the time [hours].

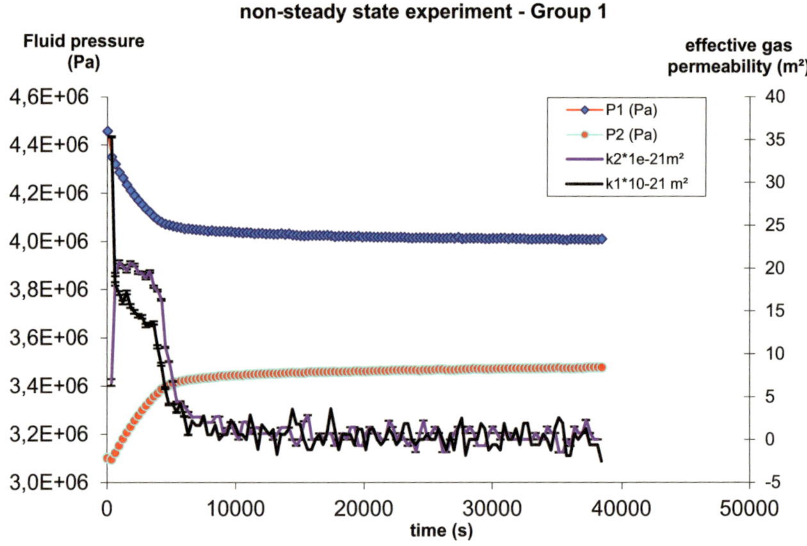

Fig. 7: Plotted values for upstream pressure P_1 and downstream pressure P_2, upstream permeability k_1 and downstream permeability k_2.

Additionally the calculation of the hydraulic conductivity (kf [m/s]) was done for this experiment. The hydraulic conductivity "of a soil is a measure of the soil's ability to transmit water when submitted to a hydraulic gradient" [2]. This just can be done for water and not for gas. The formula is a deflection of Darcy's law:

$$Q = -K \cdot A \cdot \frac{\Delta h}{\Delta x} \rightarrow K = -\frac{Q}{A} \cdot \frac{\Delta h}{\Delta L} \rightarrow k_f = \frac{(k \cdot \rho_{water} \cdot g)}{\mu_{water}} \left[\frac{m}{s}\right] \qquad (7)$$

Where g is the gravitational acceleration (m/s²), k the permeability coefficient, µ the viscosity of water and K the specific permeability (m²).

The estimated values for the hydraulic conductivity are 0 m/s for every measurement, so the average hydraulic conductivity is 0 m/s for this rock sample. A conversion into m² even let the hydraulic conductivity remain zero. This indicates that this is an impervious rock and therefore definitely not an aquifer. Permeability in this range (k<10-18m²) are typical for consolidated rocks like limestones, mudrocks, or dolomites and granites [2]. On the basis the used rock sample was a marl these values would fit.

III. Error calculation

The maximum relative error (mrl) was calculated with the following formula:

$$mrl = \left|\frac{2\Delta P_1}{P_1}\right| + \left|\frac{2\Delta P_2}{P_2}\right| + \left|\frac{\Delta A}{A}\right| + \left|\frac{\Delta(\Delta x)}{x}\right| + \left|\frac{\Delta \eta}{\eta}\right| \qquad (8)$$

For the maximum absolute error (k_1 and k_2 [mD]) this formula is used:

$$k_{1,2}\,[mD] = (mrl) \cdot k_{1,2} \qquad (9)$$

These equations for error calculation allow a statement over the maximum relative error in the used equations for k and the maximum absolute error for k_1 and k_2. The calculation has been done for every single measurement. The results are 0.011 for the maximum relative error, $1.2 \cdot 10^{-6}$ mD for the maximum absolute error k_1 and $3.4 \cdot 10^{-6}$ mD for the maximum absolute error k_2. All results represent the arithmetic mean values of these calculations.

IV. Conclusion

In the steady-state phase experiments gas breakthrough occurred for the saturated and dried samples. A higher entry pressure for the saturated sample is required than for the dried sample. This is caused by the reason that in the dried sample no capillary forces are active, because it was a single phase flow. In the saturated sample a two-phase flow occurred after gas breakthrough and therefore capillary forces are active.

For the non-steady state single-phase experiment, as mentioned above, the small gas breakthrough occurred by about 50 seconds and between 30-31 bar and a residual snap-off pressure remains.

V. References

- http://www.lek.rwth-aachen.de/cms/index.php?id=142 (28.01.13) [1]

- http://web.ead.anl.gov/resrad/datacoll/conuct.htm [2]